Breaking Your Dog's Bad Habits

Paula Kephart

CONTENTS

Introduction

Some of the bad habits that dogs develop are only mildly annoying, but others can lead to serious consequences. Whatever the habit or your level of concern, there comes a time when you realize that the habit must go. With your help, your dog can replace his bad habit with acceptable behavior in just a few weeks.

Retraining your dog does not necessarily call for a lot of extra time and effort on your part. Most retraining requires one or two 10-minute sessions a day. Once your dog has begun to figure out what you want from him, you can integrate continued retraining into the ordinary routines of your day.

What's Causing the Misbehavior?

There are a number of factors that can contribute to the development of behavioral problems. Poor health, old age, or the quirks of a particular breed may be to blame. Before you start retraining your dog, take some time to assess the factors that are contributing to your dog's bad habit. If you understand the problem, you'll find it easier to bring about a successful solution.

Is Your Dog Healthy?

Before you assume that your dog's misbehavior is willful, make sure that there is not an underlying health problem. It's not hard to imagine how a urinary tract infection, poor vision, diminished hearing, or other health problems could affect your dog's behavior. Before tackling the bad habit, it's a good idea to discuss the problem with your dog's veterinarian to rule out any health factors.

Is the Habit Connected to Age?

Some bad habits are more common at certain ages. Just think of the bad habits a puppy can get into because of his need to chew. And an "unresponsive" dog may simply be old and losing her hearing!

Puppies. Bad habits can develop if a puppy is weaned and removed from the litter too early. Watch a litter of puppies with their mother. What looks like play to us is really the process of doggie socialization. As they tumble about together, the littermates learn the limits of acceptable behavior and the importance of rank. If a puppy is removed from the litter too early, he may not have the opportunity to learn the "pack rules." Eight weeks is the youngest age at which a puppy should be taken from his mother and siblings; 10 to 12 weeks is better.

Adolescents. Dogs, like humans, go through a physical maturation process that includes adolescence. The physical and hormonal changes of adolescence, which occur between 3 and 6 years (human) of the dog's life, can sometimes undermine her ability to focus and behave appropriately. The keys to getting through this stage are patience and consistency.

Older dogs. Changes at the other end of the age continuum can affect a dog's behavior, too. Around your dog's eighth year, you may begin to notice physical evidence of the aging process. The dog moves more slowly as joints become less flexible. An older dog whose sight or hearing has diminished may be startled enough to snap at a hand reaching out to pat her. Changes in blood flow and blood-oxygen levels can impair a dog's cognitive processes, slowing her reactions.

The onset of aging is influenced by many variables, including breed, size (small dogs tend to have a longer life span than large dogs), health history, and the quality of care the dog has received. A dog that has been healthy, active, and well taken care of throughout her life may not show any evidence of age-related changes for several years after her eighth birthday.

Is Your Dog Confused about Who's in Charge?

Dogs, like people, exhibit a wide range of personalities and traits. This is what makes a dog so appealing and, at times, exasperating. A dog with an eager-to-please personality is much less likely to develop bad habits such as ignoring you or challenging your authority. On the other hand, this personality type may foster bad habits such as whining, begging for food or attention, and separation anxiety.

A dog with a confident personality doesn't demand a lot of your attention or affection. She seems to relish new experiences and adapts well to changes. Generally she is not afraid to make new friends among dogs or people. However, you may sometimes find yourself in a "who's the boss" struggle with a confident dog.

Canine Jekyll and Hyde

Some dogs have distinctive character traits that can, when activated, override their usual demeanor. Sam, my 90-pound black Lab/German shepherd mix, is a good example. Generally loving and playful by nature with both dogs and people, he loses all control when another dog "trespasses" by walking along the street in front of his home. Although 7 years old, Sam can clear the 4-foot fence around the yard with ease to challenge the "trespasser." His behavior is motivated not by viciousness but by anxiety about the safety of his family. While it may be endearing that he's willing to face his anxiety head on, the resulting behavior causes a fair amount of chaos.

Dominance factor refers to how submissive or dominant a dog tends to be with people and other dogs. His tendency toward dominance or submissiveness is not set in stone and can vary depending on the situation. For example, Gracie, my black Lab/beagle mix, is an outgoing 2-year-old. She loves people and generally runs up to — or after — anyone she sees to say hello. Yet she is very timid with other dogs and cowers when one comes along.

If your dog has a strong dominance trait, you need to be clear and consistent about your behavioral expectations and limits for him. Otherwise, your dog may get into the bad habit of taking charge, which can lead to an array of other habits, including overprotective and threatening behaviors.

Is Your Dog Outsmarting You?

If your dog is alert and quick to learn, you may find his repeated bad habit perplexing and aggravating — if he's so smart, why is he acting so badly? Although you might expect a dim-witted dog to develop bad habits, in truth it's often the opposite. A highly intelligent dog is a pleasure to train and have as a companion, but he is more likely to look for — and find — solutions to what he considers problems. For example, if he is shut in a fenced-in yard, he may figure out how to unlatch the gate to escape. Proud as the owner may be of her dog's resourcefulness, leaving the yard without permission is definitely a bad habit.

Was Your Dog Bred for Trouble?

Personality, intelligence, the dominance factor, and health are all influenced to some degree by the dog's breed. In addition, breed dogs may develop certain bad habits simply because they were bred for for certain characteristics. Beagles, for example, like to dig and whine. Labrador retrievers have a propensity for jumping or even climbing fences. Dogs bred to herd animals may exercise that tendency with the neighborhood children.

Saint Bernards make great family dogs because they were bred to be large, strong, loyal, and friendly. Without proper training, however, they can become overprotective of their owners.

Are You Giving Your Dog Mixed Messages?

One final factor that cannot be overlooked when investigating your dog's bad habit is you, the owner. Your attitude, expectations, self-confidence, and understanding of your dog have a definite influence on him. For example, if you don't give commands in a serious manner, and if you don't take the time at the onset of training to ensure and insist that your dog obey your commands, your dog may learn that he doesn't have to obey.

To be successful in retraining, it is important to clarify for yourself exactly which behaviors are acceptable and which are unacceptable. For example, do you want your dog not to jump on anyone? Or is jumping an acceptable way to greet family members, but not visitors? Do you want your dog to stay off all the furniture? Or is it okay for her to sleep on your bed? Once you've identified which behaviors you want and those you don't, it will be much easier to give a clear message to your dog.

Beginning Your Dog's Training

Retraining doesn't have to be somber business. Make the sessions fun and satisfying for you and your dog. Generally, dogs like to learn and love to perform, especially if it means earning your praise and approval. As your dog catches on to the new behavior you've deemed acceptable, she'll work harder to show you just how smart she is. You'll see her chest puff up with pride as she responds to your command correctly. And you'll puff up with pride, too, when you see your dog respond appropriately to a situation even before you give the command!

Dogs love to learn and perform, especially if it means earning your praise.

Working with Serious Behavior Problems

Serious behavior problems, such as chasing cars and being over-protective of family members, require focused retraining sessions at first. These sessions don't have to be long, but they should be free of distractions for both you and your dog. Try to fit in two or three short sessions a day initially. Once your dog understands what is expected of him, you can decrease the number of focused training sessions and reinforce the retraining during the course of your daily routines.

One of the most valuable benefits of retraining is the positive effect it has on the relationship between you and your pet. As you work together, you'll feel a kinship, a sense of being a team, that will strengthen the bond between you.

Training Techniques

There are many behavioral techniques for training animals. Learning occurs as a result of positive or negative reinforcement such as food treats, praise, or physical restraint and discipline. Many trainers believe that physical discipline (punishment) should be used as the primary method to subdue a dog. But the trade-off is that punishment creates contrariness, lack of trust, and resentment, all of which can build into aggression.

I've found that positive motivation, such as food treats or physical affection, is much easier and more effective for training than physical discipline. Furthermore, positive interaction deepens the bond of affection, trust, and respect between owner and pet.

Following are quick descriptions of the most common training techniques in use today.

Classical conditioning teaches a dog to react or respond to a signal not usually associated with the behavior. Dr. Ivan Pavlov's experiments involving training a dog to salivate at the sound of a bell are the most familiar and descriptive example of classical conditioning. First Pavlov paired the sound of a bell with an automatic response — salivating at the sight of meat. Eventually Pavlov "taught" the dog to salivate at the sound of the bell when no meat was present. Today, clicker training is a good example of classical conditioning — dogs are taught to respond to variations of the "click" from the clicker tool.

Counter conditioning teaches a dog to replace a problem behavior with a more acceptable behavior. The acceptable behavior distracts the dog and interferes with exhibition of the problem behavior. For example, if your dog begs for food from the table, you can train him to lie down at your feet or in another room during mealtime, either of which will prevent him from begging.

Desensitization is the process of gradually diminishing a dog's fear of a particular thing or event by exposing him to it in incrementally increasing amounts. This technique can be used to cure a dog's phobia of thunderstorms, for example (see the box below).

Intermittent reinforcement helps sustain the succesful training of a dog. Once he's been trained, the dog does not get a treat or reward every time he exhibits the appropriate behavior; he gets one only every now and then. He'll learn to perform the good habit consistently in the hope that *this* time he'll get the reward.

Operant conditioning is a very effective technique for "shaping" behavior through trial and error and rewards. For example, a rat learning to press a bar or a button to get food is experiencing operant conditioning. You can use operant conditioning to encourage your dog in an appropriate behavior that he sometimes performs; simply praise him lavishly or otherwise reward him every time he performs it.

Thunderstorms:
A Case Study in Desensitization

Many dogs become frightened and behave badly during thunderstorms. You can use a desensitization technique to relieve your pet's anxiety. Play a tape recording of thunder at a low sound level when the dog is relaxed or happily engaged in a favorite activity. He'll probably barely notice it. Then begin playing the tape over and over throughout the day. Gradually increase the sound level to build the dog's tolerance of it. Over the course of many weeks, the dog will eventually get used to the sound of crashing thunder, real or taped. He still may not like it, but he'll be able to tolerate it.

Rewarding Good Behavior: Bribery or a Well-Deserved Salary?

Positive reinforcement, or offering a reward for performance, helps your dog understand and then look forward to performing the behavior you want to encourage. As the dog comes to understand that a certain behavior results in something positive for him, he begins to anticipate that reward. In fact, he will learn to exhibit the desired behavior in the proper situation without your command.

There are many ways to reward your dog for appropriate behavior. You can use them individually or in any combination. They include:

- Praise
- Physical affection
- Food treats
- A favorite game, toy, or activity

Sometimes dog owners are reluctant to give their dogs any kind of reward other than praise. They feel they are bribing the dog and that instead of really obeying commands, he is just performing to get food. This is true to some extent, but all learning is driven by reinforcement, negative or positive, and positive reinforcement can be very effective. Some dog trainers compare food rewards to a dog's "salary" rather than a bribe. In other words, the dog gets "paid" for doing his job, which is to respond to your commands.

Establishing a Routine

Dogs function best in a structured and stable environment. The predictability of routines is a strong antianxiety agent. However, while routines are important, they don't have to be rigid. In fact, it's better if they are not. For example, don't try to get home at the same time each day. If your dog comes to rely on that time, she may get very anxious when traffic, weather, or some other unexpected event delays your arrival. However, you could teach your dog that your arrival at home means playtime, whether it's a few minutes romping in the house or a quick walk outdoors.

It's imperative that you be consistent in your training or, through intermittent reinforcement, you may actually encourage bad habits.

For example, the owner is almost always to blame for a dog that begs for food while people are eating. One day the dog nuzzles his owner during mealtime. The owner, in a lax moment, gives the dog a bit of food. But the next few times the dog tries this, he is rebuffed. Then another lax moment occurs, and the dog gets more food. Now the dog is not sure what to expect during mealtimes, but hope springs eternal in his mind, and he becomes a permanent fixture near the table during meals.

Giving Commands: Body Language and Voice

Most dogs are quick to learn commands, but they also respond to expectations and attitudes. As you start to retrain your dog, check your attitude. Are you giving your dog mixed messages? When giving a command, the tone and pitch of your voice, your facial expression, and your body posture should all convey a clear message. If you give a half-hearted command, you'll most likely get a half-hearted response. After all, if you're not paying attention, why should your dog?

Establishing your authority as the leader does not require brow-beating your dog into submission. Nor do you have to break his bouncy, energetic spirit. There are many effective and humane ways to communicate your authority:

- **Voice.** If you find yourself repeating a basic obedience command several times before your dog complies, make some changes in your voice. Speak clearly in a somewhat low-pitched voice. Use a confident and firm but friendly tone.
- **Eyes.** Make eye contact with your dog to let him know you are focused on him. Convey by the expression in your eyes that you expect him to comply. Raise your eyebrows in a way that says "No nonsense, buddy."
- **Posture.** Stand up straight but relaxed, the picture of confidence. Make sure your posture says that you are in charge and are confident the dog will comply.
- **Touch.** If the above techniques don't work, calmly walk over to your dog. Take his collar, give it a little shake, and repeat the command. If necessary, repeat the shake and the command again. Don't be rough when giving your dog's collar a shake. This action is not intended to be punishment. Rather, it is a gentle way to get his attention, to help him focus. Think of it as tapping someone on the shoulder.

- **Reinforcement.** As soon as your dog follows the command, reward him with verbal praise and physical affection such as scratching behind his ears or under his chin. You can even give him a food treat if you have one handy. (You may wonder why you are rewarding your dog when he did not comply immediately. The purpose is to encourage him to respond next time by making it clear that his response generates something pleasurable for both of you.)

Training Yourself to Be a Leader

A dog pack must always have a leader; otherwise chaos ensues and the pack falls apart. The leader, or alpha dog, is responsible for the pack's survival. He enforces pack rules and strives to protect the pack from harm. In a human-dog pack, who is the alpha dog? That's right — you, the owner.

For some of us, the concept of being "top dog" feels uncomfortable. We view our dogs as companions, perhaps even members of the family. We relish their affection and comfort, and we don't want to pull rank on them.

But assuming the alpha-dog role in your human-pet pack will not undermine this sense of camaraderie. In fact, it allows devotion to flourish because your dog knows he can trust you to look out for him. Remember that the dog's instinct tells him that survival depends on having a leader. If he thinks you are not up to the job, he'll become anxious. He may appoint himself as leader, a bad habit that can only lead to other bad behavior. For everyone's sake, keep this in mind: The one who buys the dog food is in charge. Your dog will thank you.

A well-behaved dog has no misconceptions about his status in the household hierarchy.

Understanding Pack Behavior

Most of us know that dogs are pack animals, but what exactly does that mean? A pack is a social structure that functions to ensure its members' survival. The pack operates by a set of rules, some instinctive, some learned. The rules help maintain order and harmony within the pack so that members can work together on essential tasks: procuring food and shelter, procreating, rearing the young, and defending against predators. The pack is organized as a hierarchy. Each member's ranking in the hierarchy defines his responsibilities and appropriate behavior toward the other pack members.

The pack structure applies to your home and family, too. You may not like to think of yourself as a member of an animal pack, but be assured that your dog sees you, other household members, and other pets as his pack. You can't change your dog's innate perspective, but you can apply the pack concepts to life with your dog, including the process of retraining.

When Your Dog Forgets His Manners

Ignoring commands is a bad habit that is often more annoying than anything else. You tell your dog to sit and he ignores you. You call the dog and he ambles off in another direction. Often these bad habits are unwittingly fostered by owners who are lax and inconsistent. Fortunately, it does not take too much effort to turn things around.

Ignores the "Sit" or "Lie Down" Command

This bad habit is exasperating. You tell the dog to sit or lie down and find yourself repeating the command several times. Meanwhile, the dog just looks at you as if you haven't said a word.

Breaking your dog of this habit is not difficult. It's mainly a process of drilling until the behavior occurs automatically at the command. In the process you will become more confident, clear, and firm in delivering the command.

You've probably already discovered that tone of voice makes a difference. The following scenario may sound familiar: After repeating a command but getting no response, you lose your patience and crossly shout the command at the dog. Miraculously, the dog finally obeys. Was it magic, or does your dog respond only to crankiness? It's not magic, but the crankiness is a clue. The dog understood what you said the first three times; he just wasn't convinced you meant it. Your cross words were spoken with conviction, and the dog finally got the message. He was testing your limits.

Follow these steps to retrain your dog to obey simple commands:

Step 1. Set aside 5 to 10 minutes twice a day for retraining your dog. During these sessions, repeat the "sit" and "lie down" commands over and over again. Reward your dog with praise, physical affection, and food treats when she complies with the command. Vary the sequence of the commands and the location of the sessions. Make them fun. Use a friendly, confident, but firm voice. Act as if you have no doubt that your dog will obey your commands.

Step 2. Begin to demand prompt compliance. Does your dog sit or lie down immediately, or does she ever so slowly lower her hindquarters or body to the ground? Start rewarding her only when she complies within three seconds. Work with her on the 3-second compliance over a few sessions.

Step 3. After your dog has mastered the 3-second compliance rule, start to reward her only when she complies immediately. There should be no delay to "think it over," however brief. The dog knows what is expected of her at the command, and she should just do it. (When my dog is feeling obstinate, he'll deliberately give a long blink or yawn before he complies.)

Step 4. Continue working on immediate compliance over several sessions until your dog responds promptly 90 percent of the time. In the meantime, keep the training interesting. For example, you can:
- Turn your back and give the commands.
- Close your eyes or put a paper bag over your head.
- Have the dog face away from you as you give the "sit" or "lie down" command.
- Give the commands from another room.

Will Not Stay until Released

How long does your dog stay when you tell him to? Does he look you in the eye and then walk away? Or does he at least wait until your back is turned before leaving the scene? If your dog has learned the basic obedience commands, it is unlikely that he has forgotten them. It's more likely that he is not taking you seriously, and probably rightly so.

Many pet owners are lax about the "stay" command. We tell the dog to stay and then get absorbed in the mail or the television or a conversation. The dog sees that our attention has wandered and takes this as his cue to wander off, too. We may notice that he has gotten up — released himself from the "stay" command — but we don't consider it a problem. After all, he isn't hurting anyone, right? Wrong. He's acting out a bad habit for both of you. Your dog is in the bad habit of not taking you, his leader, seriously. And you are in the bad habit of giving a command you are not serious about.

To retrain your dog with the "stay" command, follow these simple steps.

Step 1. Pay attention to your dog. Make eye contact with him. In a calm, confident voice, give him the "stay" command. If he complies immediately, praise him. If he does not comply immediately, wait half a minute, then repeat the command using his name: "Sam, stay."

Step 2. If he ignores the "stay" command and starts to walk away, stop him by gently taking hold of his collar. Calmly and cheerfully repeat the "stay" command and let go of the collar. If he does not stay, repeat the process until he does. When he is obedient, praise him with "Good stay," but not too enthusiastically, or he may assume he has been released.

Step 3. With your dog in stay position, turn your attention to something else, but keep an eye on the dog. If he starts to leave, calmly remind him to stay.

Step 4. Release the dog. Make eye contact, speak clearly, and use his name with the release command: "All right, Sam." Praise him for a good stay. Give him a treat if you wish. Always focus on your dog when you give the release command. This distinguishes the "stay" command for both of you as having a beginning and an end.

Step 5. When your dog has mastered the "stay" command while you remain nearby, it's time to practice leaving him behind with the "stay" command. Repeat the "stay" command and walk away from the dog. If he starts to move, calmly remind him to stay. Walk away — whether across the yard or to another room — but not too far away. Wait a few seconds or minutes, depending on how much time you have, before giving the release command. You want him to know the release was your idea, not his.

When you tell your dog to stay, he should do so until you give the release command. You should tell your dog to stay only when you really mean it. Practice long and short stays and leaving the room after giving the "stay" command. You want your dog to obey the "stay" command regardless of the circumstances.

Repeat these steps several times a day over the course of a couple of weeks. Integrate training sessions into your daily activities. For example, you can have your dog practice "stay" while you're doing housework, helping the kids with their homework, brushing your teeth, or mowing the lawn.

Will Not Come When Called

There is nothing more aggravating than calling your dog and getting no response. The dog acts as if she didn't hear you, or she decides your call is the kickoff to a game of tag — and you're it.

Not responding to the command to come is such a common problem that many dog owners view it as a way of life. But it doesn't have to be, nor should it be, inevitable. You need your dog to come when called no matter what the reason or what distractions surround her.

Practice the following retraining techniques in different locations, including unfamiliar ones, and with varying levels of distractions. Continue the process for as many days or weeks as necessary, until you can depend on your dog to come promptly on command at least 90 percent of the time.

Step 1. Put the dog on a long or retractable leash in a large area with minimal distractions. Have her sit or stay. Holding the leash, walk a few feet away from her and say "Come!"

Step 2. If she comes, reward her with verbal praise and a treat, if you wish. If she does not come, gently tug on the leash to pull her to you. When she reaches you, pet her and say "Good come."

Step 3. Have the dog sit or stay. Walk farther away from her than before. Say "Come!" and follow step 2.

Step 4. Repeat this process for about 10 minutes once or twice a day for a couple of weeks. After a few days of success, continue the retraining in a place with distractions — children playing, traffic going by, other dogs. Follow the steps above until your dog reliably comes on command despite distractions.

Step 5. Once your dog comes reliably on command among distractions, remove the leash and go through the above steps again. If at any time your dog does not come when called, put the leash back on, say "Come," gently pull her to you, then remove the leash and try again.

Teaching Habits That Benefit You

When replacing your dog's bad habit with a more appropriate behavior, think about turning the habit around to your benefit. For example, Gracie, my black Lab/beagle mix, has a favorite morning game. She grabs my slipper so that I will give chase. I decided to exploit her game. When Gracie took off with my slipper, I would use the phrase "Bring the slipper to me." As soon as she did so, I rewarded her heartily with praise, affection, and a food treat.

Now I can tell Gracie to bring the slipper to me even if the slipper is not in the room. Her beagle nature responds to the challenge of the hunt, and she soon returns with the prize. Then I tell her to "go get the other slipper." Off she goes again. The successful retrieval of the first slipper is rewarded with praise; the second slipper warrants a food treat — the surest way to Gracie's heart.

When Your Dog Acts Out

It's hard to relax around a dog that acts out. You open the door to go outside and the dog bolts past you, knocking you into the door frame. When company visits, the dog is restless and unmanageable. These moderately bad habits are often attention-getting behaviors, but the kind of attention they attract brings little pleasure to you or your dog. Worse, such behaviors tend to undermine your confidence in your ability to manage the dog. Try the techniques in this section to regain control.

Jumps on Family Members or Visitors

One of a dog owner's greatest pleasures at the end of a long, tiring day is to be greeted with loving enthusiasm by her dog. A dog's exuberant delight in seeing us warms our heart, eases our tension, and puts life back into perspective. It's unacceptable, however, for a dog to demonstrate his affection by jumping up on you and wildly pawing at you. It's even worse when the dog displays this behavior indiscriminately to everyone who comes through the door — family members, friends, even strangers.

A dog who expresses his enthusiasm in greeting people by jumping up on them is physically out of control. The habit is not only irritating but also downright dangerous. Unwary visitors are at risk of being scratched or knocked down. At the very least, the dog's behavior puts clothing at risk for tears and dirt stains.

So how do you keep the affection without the damages? Breaking this habit is not difficult, but it takes some planning and focused training time — 10 to 15 minutes once or twice a day.

Step 1. Attach your dog to a lead and leave him on his own for a few minutes. If you're outside, attach the lead to a post or a spike in the ground. Inside, secure the dog to something sturdy, such as the leg of a heavy piece of furniture.

Step 2. Return a few minutes later, with some treats in your pocket if you wish, and approach the dog to pet him. When he starts to jump up and paw you, step back just out of his reach. In a calm, friendly, but firm voice, give the "sit" command coupled with a new phrase. For example, say "Sit to say hello." (The "sit" command tells your dog what the appropriate behavior is. Adding a word such as *hello* helps him connect the appropriate behavior to greeting situations.) Repeat the command several times, if necessary, until he obeys. Remember to keep your tone friendly and confident. This is not a punishment, but rather like teaching a child when to say please and thank you.

Step 3. As soon as your dog sits, step forward to pet him. If he remains in a sitting position, pet and praise him warmly. Incorporate your greeting command in the praise: "Good sit to say hello, Sam!" Naming the behavior with the praise is a reinforcement. If you're using treats, give one to the dog now. If the dog jumps up again when you step forward to pet him, step back again just out of his reach. Repeat steps 1, 2, and 3.

Step 4. Repeat this training procedure once or twice a day for a couple of weeks until it is clear that your dog understands what behavior your greeting command demands. As your dog begins to make the association, use the command at appropriate times during the normal comings and goings of your day.

Step 5. When your dog has learned how to welcome you happily yet appropriately, it's time to practice the command with visitors. Put the dog on the lead and have a friend walk up to the dog to say hello. If the dog starts to jump up, your friend should step back just out of reach and give the "sit" command coupled with the greeting phrase. As soon as the dog complies, your friend should reward him with petting, praise, and a treat. Repeat this routine several times in the first session, and arrange several more practice sessions over the next week or so. You can even use different friends for different sessions.

Step 6. Once your dog demonstrates successful behavior throughout a couple of sessions, have a friend come to the door and knock. Put your dog on the secured lead. Tell him to "sit to say hello" before you answer the door, and repeat the command as your friend enters. The friend should greet your dog only when the dog is sitting. If the dog starts to jump up, your friend should step back. Repeat this procedure until the dog sits while the friend greets him. Again, praise, petting, and a treat will reinforce the good behavior.

Step 7. Once your dog demonstrates compliance with visitors at the door, practice the same behavior without the lead attached.

Step 8. After your dog becomes reliable in greeting visitors without jumping, use intermittent reinforcement (see page 8) with a food treat. The dog's anticipation of a treat will encourage him to maintain appropriate welcoming behavior.

Be Patient

Don't expect perfect performance in the first few training sessions, especially if your dog has been in the habit of jumping and pawing for quite a while. Reinforce even a few seconds of compliance with praise, petting, and a food treat.

Begs for Food When People Are Eating

There are a couple of different ways to retrain a dog who has developed the habit of begging during your mealtimes. One way is to restructure the environment. In other words, don't allow the dog in the room while you are eating. But many people like to have a snack while they watch television or read. What should you do with the dog then?

The "down-stay" command is effective retraining for begging. Your dog is probably already familiar with this command or some variation of it.

Step 1. As you sit down to eat, point to the spot where you want the dog to be and give her the "down-stay" command (or your version of it). Your dog should lie down and not get up again until you give her the release command.

Step 2. If your dog does not stay down, put her on a leash. Have her lie down on the floor next to you, and step on the leash throughout the meal. Make sure there is enough slack for the dog to rest comfortably, but not enough to allow her to sit or stand up. If your dog tries to get up, repeat the "down-stay" command.

You can use other techniques to limit your dog's begging. Perhaps you don't mind sharing a morsel or two with your dog, but you want it to be at your discretion, not hers. If you're having a snack and have no intention of sharing it with your dog, say "Not for dogs" and don't give her even a sliver. You want her to learn that this phrase means absolutely no chance. When you are willing to share a small amount, use the phrase "That's all." For example, after giving the dog one or two potato chips, say "That's all," and do not give her any more. You'll find that she quickly learns and accepts these limits.

Remember, a dog begs at the table most often because her owner is neither clear nor consistent. Feeding your dog from the table occasionally is the same as training her to beg from the table using intermittent reinforcement. When she sees people eating, she won't be sure she'll get a treat, but she'll know that she *may* get a treat, and so of course she'll come and beg, hoping that this will be her lucky day.

Breaking Dangerous Habits

If your dog is exhibiting any of the bad habits in this section, take them seriously. Such behaviors go beyond mere attention getting or acting out — they indicate that your dog has assumed the alpha-dog position in the household pack. In doing so, he puts household members, and possibly outsiders as well, at risk. Use the following techniques to take control again, but do so graciously.

Displays Jealousy or Overprotectiveness

Picture this: a candlelight dinner, soft music in the background, and you and your favorite person alone at last. You lean forward for a kiss and hear a low, throaty growl. No, it's not your dinner partner. It's the dog. He's staring at you with a menacing glower that quickly chills the romantic moment.

It is not uncommon for a dog to become jealous of his owner's significant other. If so, he'll treat that person as an interloper. The dog views anyone who is part of the household as a member of his pack; anyone else is a potential interloper. If he accepts his owner as the alpha dog but considers himself second in command, he is likely to display possessiveness or overprotectiveness of his owner.

If the dog's behavior is not recognized as a bad habit and goes uncorrected, it will gradually escalate from obnoxious to intimidating. The dog may stare, growl, bare his teeth, snap, or even bite the interloper.

An Unwanted Escort

If a dog is jealous, he is usually far from subtle in making his feelings clear. At first he may simply rest his head on your knee when the "interloper" is nearby. Or he may nudge his way in between a hug. Some couples report that the dog will not allow them to hold hands, hug, or kiss in his presence. Some dogs have even gone so far as to try to keep the interloper out of the bed or bedroom! At this point, the dog's bad habit is clearly out of control.

As the owner, you may initially find your dog's behavior endearing. But your loved one is not likely to be pleased with playing second fiddle to the dog.

Sometimes a dog who thinks he's in charge will focus his possessiveness on the children in the household. Their small size, high-pitched voices, and uncoordinated movements signify that they are the pack's young and therefore defenseless, so the dog appoints himself their bodyguard. As with possessiveness of an owner, bodyguard behavior is a bad habit with serious consequences. For example, the dog may think someone is acting in a threatening manner toward one of the kids and will therefore act to protect her. He may snap at, bite, or assault the offender, even if the person in question is actually a harmless parent or playmate.

To break your dog's bad habit of jealousy and possessiveness, you must restructure the hierarchy of the household pack. In the new order, the owner is the alpha dog and any other humans, regardless of age or size, are second in command. Dogs are last in the new hierarchy. The new order must be made clear to the dog in a nonconfrontational way.

Everyone in the household must abide by and reinforce the rules and expectations for the dog's behavior. No mixed messages! The owner should use her alpha position to support the other household members' interactions with the dog. The owner and all human members of the household must handle all interactions with the dog with calm confidence, firm but nonconfrontational. This may be hard to do at first if the dog has growled or snapped at anyone, so fake it until it becomes natural.

The following steps can be performed with the assistance of whomever the dog considers to be an interloper, whether a particular significant other, a baby-sitter, or a friend of the household.

Step 1. You and the person your dog perceives as the interloper should arrange one or two daily retraining sessions with the dog. Make the sessions brief — 10 minutes is plenty. During these sessions you will begin to share your authority as alpha dog with the other person. For the first few sessions, you should give the dog simple obedience commands — sit, stay, lie down, come. When the dog responds appropriately, praise him warmly and pet him. The other person should join in with warm praise, too. However, this person should not give any commands or try to reward the dog with petting at this time. After all, he or she is not the dog's favorite person right now.

Step 2. After a few sessions, the other person should take over the retraining sessions. The owner should be on hand for support. Again, the sessions should be brief and nonconfrontational and should cover just the basic commands. The dog should be rewarded with praise and food treats. (Food will serve as a strong motivator to get the dog to accept this person's commands.)

Step 3. Once the dog is comfortable interacting with this person, the owner should not attend the practice training sessions. Gradually, over a few weeks, the dog will learn to accept the other person's authority in this and other situations. Remember that the retraining sessions should be fun and relaxing for both humans and the dog. Interactions with the dog outside of retraining sessions should remain confident and friendly but nonconfrontational. For example, if the dog is doing something he should not or ignores a command, don't try to force him. Instead, distract him with another activity or a toy.

Pulls on Leash When Walking

We've all seen or experienced this scenario: The leash is stretched as far as it will go, and the dog is straining against it as if he is pulling a heavy sled in the Iditarod. The owner is leaning backward in an effort to slow the dog or, worse yet, is being dragged along like a loose anchor. Such scenes make us wonder just who's walking whom!

Amusing though the scene may be, pulling on the leash is a serious behavior problem for several reasons. A dog that pulls his owner along is either very anxious or is vying for the alpha-dog position. Either way, if the dog is in control of the walk, the owner is not in control of the dog. Many potential problems loom in such situations. The dog may suddenly lunge at a passing bicycle, runner, or child, or the dog may go after another dog or a cat. Serious injury could result to any or all involved. In addition, the constant tugging of a determined dog on a leash can cause stress injuries to the owner's arm muscles and joints. There is even some possibility that the pulling could harm the dog's neck or throat.

Taking control of walks is not always easy for the owner. Neither brute force nor begging is effective. To correct this bad habit, try a change in both "hardware" and technique. Use a collar or harness that minimizes a dog's ability to pull — without choking him — and retrain your dog to walk with you. Here's how.

Step 1. Several types of collars and harnesses can minimize or eliminate pulling, and they are more effective than jerking on the dog's leash and more humane than choke or shock collars. Do some research. Discuss the pros and cons of various types of collars and harnesses with people who have had some experience with them — your vet, a dog behaviorist, other dog owners. Get their opinions on what works and why. Every dog is different, so what works well for one won't necessarily work for another.

After you've done your research, select the type of equipment that you feel is most suitable for your dog (see pages 24–25 for suggestions). Consider your dog's size and physique, personality and temperament, and dominance factor. Take your own feelings into consideration, too. If you feel that a particular collar or harness is uncomfortable or undignified for your dog, you'll communicate that distress right down the leash.

Step 2. When you first bring home the new collar or harness, don't try it out right away. Put it down on the floor and let your dog check it out. As he does so, talk to him. Explain that this is something special for him, something to make walks easier for both of you. Use a cheerful tone. Don't scold the dog or give the impression that the new item is a punishment. It isn't. It's a tool that will help you and your dog feel calm and confident so you can enjoy your walks.

Step 3. After he's had a minute to sniff the new item, offer your dog a treat. Set it down on the floor next to the harness or collar to help him associate good things with the new equipment right from the start.

Step 4. Find a quiet time to try the new collar or harness on the dog. Use a casual, "nothing-out-of-the-ordinary" manner. Be sure to have several tidbits on hand to ease the experiment. If you are using a Halti or Gentle Leader collar that slides over the dog's snout, put a tidbit in the palm of your hand so that the treat will slide into the dog's mouth as the collar goes on.

Step 5. Talk to your dog calmly and quietly. Tell him how good he is. Remind him that this new collar or harness is going to make walks more fun. He'll most likely resist the device at first. Leave it on for a few minutes, then gently take it off and leave it on the floor. Praise your dog and pet him. Let him know you are proud of him. Offer him a treat from your hand, then set a second treat down on the floor near the collar or harness.

Step 6. Later that day or the next, announce walk time. Be sure to put several treat tidbits in your pocket. Calmly, without a fuss, slip the collar or harness onto the dog and in the same moment give him a treat. Spend a few minutes practicing walking back and forth near your home. Don't head off for a real walk this time.

Step 7. If the dog starts to pull on the leash, correct the movement according to the directions that came with the harness or collar. Tell him "No pulling," but use a gentle, calm, and encouraging voice. You don't want to baby him, but it is important to give him some time to get used to the change.

Step 8. Once you're back home, remove his new equipment and give him praise or a treat.

Step 9. When it's time for a walk again, plan a short outing of 10 to 15 minutes, just long enough for both of you to start getting used to the new equipment. Lead off gently. If need be, use an encouraging tone to prompt your dog to walk. If he is reluctant to do so, coax him along

---

Tools for Restraining Your Dog during Walks

These items can be purchased from pet stores and animal supply catalogs and Web sites. Some veterinarians, dog trainers, and dog behaviorists sell them as well.

Equipment	How It Works
Head collar	Collar loops over dog's snout and back of head. Leash attaches to a ring under the dog's chin. A gentle pull on the leash exerts a light pull on the dog's head.
No-tug or no-pull harness	Fits around the dog's chest and legs. Some apply pressure to the chest, some to the front legs, some to the back legs. The pressure stops the dog from straining on the leash.
Sound-emitting leash attachment	Small, battery-operated device that attaches to the dog's leash. When the dog pulls on the leash, the device emits a high-pitched, shrill tone that distracts the dog and interrupts pulling behavior.

with a tidbit or two. He'll get the idea quickly, though he may resist the new device at first. Don't relent: No collar or harness = no walk. Once they get moving, most dogs get absorbed in the walk and forget about the harness or collar. If the dog pulls, correct him immediately using the proper technique for that device. Say "No pulling" in a firm, but not angry, voice. When he starts walking properly again, praise him for doing a good job and give him a treat. When you return home from the walk, reward him for a "good walk" with praise and physical affection, and give him a treat *before* you remove the collar or harness.

Step 10. For your next walk and thereafter, follow your usual route. Your dog will learn quickly, within two or three walks, to walk without pulling, but it may take a few weeks before the new behavior is consistent. For the first month, reward your dog a few times on every walk when he walks properly. Pat him on the head and slip him a treat. Gradually, he'll begin to take pride in his new behavior.

Pros and Cons	Brand Names	Price Range
Makes it easy to control a big or strong dog without much physical exertion. Some dogs cannot abide having something around their snout or mouth, even though it does not inhibit breathing or panting.	Halti, Gentle Leader	$10 and up
Helps the owner keep control of the dog. Some dogs resist or feel anxious with this kind of confinement.	Top Paw Holt, Pro-Stop!	$10 and up
Tone reinforces verbal commands. More humane than electric shock devices. Some dogs can ignore the tone; some owners cannot.	Happy Walker	$20 and up

Persuading a Stubborn Dog to Walk

What if your dog absolutely refuses to walk after you've put on the new harness or collar? If she is small enough, pick her up and start walking. Put her down when you are a few yards away from your home, give her a treat, and start walking again. This tactic is usually enough to communicate to the dog that the collar or harness comes with the walk.

If your dog is too big to carry, spend a few minutes coaxing her to walk. If she still refuses, take her back into the house and take off the harness or collar without comment. If time permits, try again in another 15 minutes or so. Try this three times before you give up. If you strike out after three tries, put the harness or collar on your dog but don't use them. Instead, attach the leash to her regular collar and take her for a short walk that way. This will let her get used to walking while wearing the device, but without any pressure. Since the leash is not attached to the restraining device, you will not be able to physically correct pulling, so keep these walks short.

After three walks in this manner, attach the leash to the restraining device and start out for a walk. By this time, she will be used to the feel of the new harness or collar. Now you can start working on correcting the pulling behavior. Be sure to praise and reward your dog when she walks properly.

Chases Cars

Cubby is such a feisty 5-year-old Pomeranian that it's hard to believe she nearly didn't make it to her first birthday. When she was just 6 months old, she was severely injured by a car. Cubby and her owners live on a side road off a busy country road. Because of the heavy traffic on the main road, her owners always put Cubby on a leash whenever she is outside. One evening, Cubby's owner unclipped her leash just as they were walking up the front steps. At that moment a neighbor's car turned onto the side road in front of their house. The crunch of tires on gravel startled Cubby, and she spun around to defend herself and instinctively went on the attack. She hurtled herself toward the car. The car's front tire clipped Cubby and sent her rolling along the gravel, head over tail.

Cubby suffered extensive internal damage and severe abdominal lacerations from the accident. As a result of the accident she becomes enraged with any vehicle other than her owner's that passes the house. Whenever she hears or sees a car, she breaks into frenzied barking. Even if she is inside the house when a vehicle is passing, she reacts with fury, racing around in the hope of catching the offending car.

Cubby's accident and the behavior that resulted from it vividly portray the dangers for dogs who chase cars. But the behavior puts humans at risk, too. A dog hurtling out of nowhere can startle a driver enough to make him go off the road or into oncoming traffic. And many people are devastated when they accidentally hit an animal.

Desensitization is an excellent technique to use for banishing car-chasing behavior. Desensitization works by gradually exposing the dog to the stimulus that provokes his terror or aggression so that over time his reaction (sensitivity or overreaction) to the trigger will gradually decrease.

Throughout the desensitization process, try to avoid any situations that will trigger your dog's old fearful or enraged reaction. If such an incident does occur, it can undo the desensitization you have accomplished, and you will need to start the process all over again. You should plan on several training sessions a week over many weeks. During the sessions, keep your dog on a leash and in your control at all times.

Laying the Groundwork

For desensitization to work, it is important that your dog has mastered basic commands such as *sit*, *stay*, and *lie down*. If he does not promptly obey basic commands, go through the retraining steps for that bad habit first. Once he responds consistently, you'll both be ready to tackle more serious behaviors.

Step 1. Study your dog's behavior in response to vehicles (the stimuli). You may see some subtle clues before the dog decides to bolt. When does he start to react — when he hears a car or when he first sees it? Or does he seem to know that it is coming before you see or hear it? Once you know the clues, start trying to figure what your dog is feeling. Is it fear, anxiety, or anger?

Step 2. Identify the treat or reward that most pleases your pet. What is his absolute favorite treat, the thing that makes him wriggle with joy? Whether it's a food treat, a toy, or even a favorite game you play with him, you'll use it regularly as part of the desensitization process.

Step 3. Take your dog to an area where cars will pass by occasionally, but avoid heavily trafficked areas. Your goal is to enable your dog to be aware when a vehicle is passing but to be far enough away from the vehicle that it does not trigger the dog's usual out-of-control reaction. (This is where your knowledge of how and when the trigger affects your dog pays off. If he is first alerted by the *sound* of an approaching vehicle, you want to position him so that he will just barely hear it. If *sighting* a vehicle is the trigger, have him close enough to see a vehicle pass but far enough away that he does not react to the sight.) You'll know that you are too close if your dog starts to get hyper as vehicles approach.

It is important for *you* to know ahead of time when a vehicle is approaching so you can be ready with the treat. Position yourself where you can see an approaching car before your dog senses it, or recruit a friend to stand closer to the traffic and act as lookout.

Step 4. As soon as your dog notices the approaching vehicle, engage him with his favorite treat. This will start the process of associating a pleasurable experience with the negative one. Talk to him calmly (about anything) in a cheerful, confident voice, then walk in the opposite direction, putting distance between your dog and the vehicle to decrease any threat your dog may feel. Once the vehicle has passed, go back to where you were before and do the same thing over again several times.

Step 5. Repeat this process several times in each training session, and make time for several training sessions a week for several weeks. When it is evident that your dog feels nothing but anticipatory pleasure (the expectation of a treat) at the sight or sound of a vehicle approaching, you'll know that the pairing process has taken place. In other words, your dog has begun to associate pleasure — not anxiety, fear, or pain — with an approaching car.

Step 6. Move your training sessions a little closer to the road. Again, your goal is to have your dog far enough from a passing vehicle that he does not react to it, but close enough to be aware of its approach. As before, position yourself or a friend to be aware of an approaching vehicle before your dog is. Repeat step 4 in the new location.

Step 7. Continue gradually moving closer to passing vehicles. Don't try to hurry things along by moving too close too soon, however. End this phase of desensitization when you have reached a point of about 10 feet from an approaching vehicle. Continue "rewarding" your dog as soon as he knows a vehicle is approaching, then turn and walk him away from the vehicle while talking to him. Repeat until your dog can successfully tolerate the approaching car while staying focused on you and his treat.

Step 8. The next phase is to reverse the order of the reward and the stimulus. Because of the previous desensitization sessions, your dog should not show any fear or anxiety to approaching vehicles. Now hold off giving your dog his treat until *after* the vehicle has passed. The purpose of this shift in timing is to train your dog that the thing he feared most — an approaching or passing car — indicates that one of his favorite treats is on the way. Once your dog makes that connection, he should rarely react to an approaching vehicle with anything but gladness.

Why Do Dogs Chase Cars?

There are several theories about why dogs chase cars. A popular one is that dogs view cars as large, threatening animals and go on the attack. I disagree. Dogs are certainly intelligent enough to distinguish between machines and animals. While some dogs fear vehicles, most dogs comprehend what vehicles are for and delight in car rides.

It seems likely that car chasing is an instinctive reaction in dogs who have the innate drive to chase or herd anything that moves. However, there does seem to be an element of viciousness — ferocious barking and bared teeth — present when a dog chases a car. I suspect that tires on a road surface make a noise that, although inaudible to the human ear, is intolerable to a canine's sensitive hearing. This would explain the anxiety, fear, or rage that is often a component of car chasing.

Whatever the reason, the behavior is serious. It goes without saying that any dog who chases cars should always be kept behind a fence or on a leash or lead whenever he is outside. To allow a dog with this bad habit to run free is asking for serious trouble and heartbreak.

When All Else Fails: Calling In the Experts

Sometimes your dog's bad habit is just too serious or entrenched for you to solve the problem on your own. The habit may be rooted in emotional trauma the dog experienced, or a neurological or chemical imbalance may be causing the behavior. (Compulsive-obsessive behaviors such as shadow chasing and fly catching are examples.) Such behaviors can cause serious impairment in the dog, so prompt professional evaluation is crucial.

Professionals who can help you with retraining include veterinarians, dog trainers, and animal behaviorists. The first place to start when you need help with your animal is to ask for recommendations from dog owners you trust. Don't overlook local animal shelters and pet shops as sources of recommendations, too. If you blindly select a professional from the telephone directory, ask for and check references. Arrange a telephone interview with the professional to help you determine whether your dog-training philosophies and styles are similar.

What to Look For in an Expert

Education and professional credentials are usually good indicators of a professional's expertise. However, expertise is useless if you or your dog dislikes or distrusts the expert. Trust your instincts — and your dog's. If either of you does not feel comfortable with this person, find someone else. Never follow someone's advice, "expert" or not, if it seems cruel, harsh, or traumatic. The only thing your dog will learn from such treatment is that you can't be trusted with his welfare.

Trainers. There are various types of trainers, from those who teach basic obedience to those who train dogs for specific tasks or jobs. For basic obedience, the trainer's job is to teach you, the owner. A good trainer can get just about any dog to comply with new commands almost immediately. But that does you no good once the trainer is gone; the trainer needs to teach you the same techniques. However, don't expect to immediately get the same flawless compliance from your dog that the trainer does. You and your dog will need some practice before you can both get it right.

Animal behaviorists. Animal behaviorists can have different levels of education. Some of these professionals specialize in a specific type of animal, for example, dogs, dolphins, or horses. The dog behaviorist I consult with has a Ph.D. in animal behavior. Her consultations include:

- A comprehensive review of the dog's history, environment, and behaviors
- A 2-hour appointment with dog and owner in their home
- Demonstrations of appropriate training techniques
- Written recommendations and follow-up phone calls to check on progress

Useful Web Sites

Dogwise
www.dogwise.com
An online bookstore with books on every topic about dogs.

Fido Friendly
www.fidofriendly.com
A lifetime magazine for dogs and dogowners — includes travel tips and an apartment finder.

Karen Pryor, clickertraining.com
www.clickertraining.com
Information about and products for clicker training for animals.

PetStore.com
www.petstore.com
An online pet store with products for and information about dogs and other pets.

Pet Sitters International
www.petsit.com
Locate a local petsitter.

Other Storey Books You Will Enjoy

The Dog Behavior Answer Book by Arden Moore
View the world through your dog's eyes with the help of this friendly handbook. Dog expert Arden Moore draws from years of professional experience to answer all your questions about canine quirks, baffling habits, and destructive behavior.

Dr. Kidd's Guide to Herbal Dog Care
by Randy Kidd, DVM, PhD
Maintain your dog's health the natural way with gentle, chemical-free treatments for health problems and preventative care. These herbal remedies and health care tips can help keep dogs of any age and breed happy and active throughout their lives.

A Kid's Guide to Dogs by Arden Moore
Start dog ownership off right! With fun-to-read facts about dog breeds, understanding dog body language, training tips, healthcare how-tos, and DIY dog treats and toys, this guide helps pave the way to a forever friendship between kids and their canines.

A Kid's Guide to Cats by Arden Moore
Kids can play an active role in cat care with this fun-filled guide written just for them. From understanding feline behavior to tips for keeping cats healthy, DIY projects for kitty enrichment, and more, it's all paws on deck!

Real Food for Dogs by Arden Moore
These 50 well-balanced, vet-approved recipes use human food as a basis for delicious meals for dogs. Please your canine's gastronome and laugh out loud at the delightful illustrations as you cook.

Join the conversation. Share your experience with this book, learn more about Storey Publishing's authors, and read original essays and book excerpts at storey.com. Look for our books wherever quality books are sold or by calling 800-441-5700.